Y0-DBU-739

DODD, MEAD WONDERS BOOKS

Wonders of the Weather by Tex Antoine
Wonders of the Mosquito World by Phil Ault
Wonders of Animal Migration by Jacquelyn Berrill
Wonders of Animal Nurseries by Jacquelyn Berrill
Wonders of the Antarctic by Jacquelyn Berrill
Wonders of the Monkey World by Jacquelyn Berrill
Wonders of the Arctic by Jacquelyn Berrill
Wonders of the Fields and Ponds at Night by Jacquelyn Berrill
Wonders of the Seashore by Jacquelyn Berrill
Wonders of the Woods and Desert at Night by Jacquelyn Berrill
Wonders of the World of Wolves by Jacquelyn Berrill
Wonders Inside You by Margaret Cosgrove
Wonders of the Tree World by Margaret Cosgrove
Wonders Under a Microscope by Margaret Cosgrove
Wonders of Your Senses by Margaret Cosgrove
Wonders of the Reptile World by Helen Gere Cruickshank
Wonders of the Rivers by Virginia S. Eifert
Wonders Beyond the Solar System by Rocco Feravolo
Wonders of Gravity by Rocco Feravolo
Wonders of Mathematics by Rocco Feravolo
Wonders of Sound by Rocco Feravolo
Wonders of the Heavens by Kenneth Heuer
Wonders of the World of Shells by Morris K. Jacobson and William K. Emerson
Wonders of Heat and Light by Owen S. Lieberg
Wonders of Magnets and Magnetism by Owen S. Lieberg
Wonders of Animal Architecture by Sigmund A. Lavine
Wonders of Animal Disguises by Sigmund A. Lavine
Wonders of the Anthill by Sigmund A. Lavine
Wonders of the Bat World by Sigmund A. Lavine
Wonders of the Beetle World by Sigmund A. Lavine
Wonders of the Fly World by Sigmund A. Lavine
Wonders of the Hawk World by Sigmund A. Lavine
Wonders of the Hive by Sigmund A. Lavine
Wonders of the World of Horses by Sigmund A. Lavine and Brigid Casey
Wonders of the Owl World by Sigmund A. Lavine
Wonders of the Spider World by Sigmund A. Lavine
Wonders of the Wasp's Nest by Sigmund A. Lavine
Wonders of the Dinosaur World by William H. Matthews III
Wonders of Fossils by William H. Matthews III
Wonders of Sand by Christie McFall
Wonders of Snow and Ice by Christie McFall
Wonders of Stones by Christie McFall
Wonders of Gems by Richard M. Pearl
Wonders of Rocks and Minerals by Richard M. Pearl
Wonders of the Butterfly World by Hilda Simon
Wonders of Hummingbirds by Hilda Simon
Wonders of Our National Parks by Peter Thomson
Wonders of Flight by Robert Wells

A. DATTILIO
'73

Wonders of BARNACLES

Arnold Ross and William K. Emerson

Illustrated with photographs and drawings

DODD, MEAD & COMPANY - NEW YORK

Frontispiece:

The plants and animals comprising the wharf pile community of southern California waters include acorn barnacles and stalked barnacles, together with such predators of barnacles as starfish, annelid worms, and snails.

PICTURE CREDITS

American Museum of Natural History, 10, 13, 35; Anthony D'Attilio, frontispiece, 11, 14, 17 (left), 21, 25, 29 (right); Dallas Clites, 48; courtesy of C. L. Hubbs, 47; Michimaru Inaba, 26; Arthur S. Merrill, 59; David Mulliner, 33; John Nordback, 50; Larry Ritchie, 30; Arnold Ross, 8, 16, 17 (right), 19, 23, 32, 36, 39 (right), 41, 57, 63, 64; Edwina Stone, 39 (left); A. E. Verrill, 43.

ISBN: 0–396–06971–1
Library of Congress Catalog Card Number: 74–3783
Printed in the United States of America

For Cecelia

Contents

1

Wonders of Barnacles

Barnacles are one of the wonders of the animal world. They may seem to be peculiar or bizarre creatures, especially when they are compared with more familiar kinds of invertebrates—animals without backbones. Yet they are among the most successful denizens of the seven seas. There are probably fewer than one thousand different kinds, but some of these occur in such great profusion on rocky shores around the world that scientists often refer to the present time as the "Age of Barnacles."

Anyone who has explored rocky seashores has found acorn barnacles. These creatures resemble small volcanoes or tepees, and their limy, that is, calcareous shells can be found even on rocks in the splash zone, which is reached only by the highest tides. Acorn barnacles may cover virtually every rocky surface that is exposed to the sea. They are so common in some areas that more than two billion of them may live in a zone one yard wide along a mile of rocky shore. Yet most kinds of barnacles are not found along the shore, for many live between the shoreline and the deep floor of the oceans. And they are to be found

The stalked barnacles (Pollicipes polymerus), *in the center of the picture, are growing on a wooden wharf piling in San Diego, California. This is the only stalked barnacle that has adapted to living in the intertidal zone.*

Acorn barnacles (Balanus crenatus) *cover most of the surface of rocks low in the intertidal zone near Nahant, Massachusetts.*

at all depths between the polar regions and the tropics.

Barnacles can permanently cement or attach themselves to all sorts of things. The fixed-life habit of these creatures evolved more than 400 million years ago, long before the dinosaurs appeared, and they have survived while the dinosaurs and many other animals have become extinct. During their juvenile stages, barnacles are free to swim or drift about with the currents before settling down. It is not unusual to find barnacles on the hulls of ships, wharf pilings, sea walls, floating logs, and even on fish, turtles, whales, and many kinds of shellfish. Some, like the goose barnacles, may have a foot-long muscular stalk by which they are attached to some supporting object. Others may become hidden as parasites in the interior or on the surface of starfish, corals, and crabs. Some burrow into sea shells and corals. Although barnacles live in a variety of unusual places, it is obvious that all barnacles are confined to rather specialized habitats. What could be more specialized than the skin of a whale, or for that matter the splash zone of a rocky shore?

The shell of a barnacle is not spectacularly colored like many sea shells. Most barnacles have ash-white shells, but others may be pale yellow, pink, blue, red, or purple, and some have rayed or striped color patterns. Most barnacles are relatively small. Some are less than ⅛ of an inch, but most commonly they are about ½ inch in diameter. Others reach gigantic proportions. One kind of acorn barnacle that lives along the coast of Oregon and Washington attains a weight of more than 1½ pounds, and grows to a height of five or six inches and a diameter of four inches. Its empty shells are often used as paperweights and pencil holders.

Barnacles were assumed to be mollusks until well into the

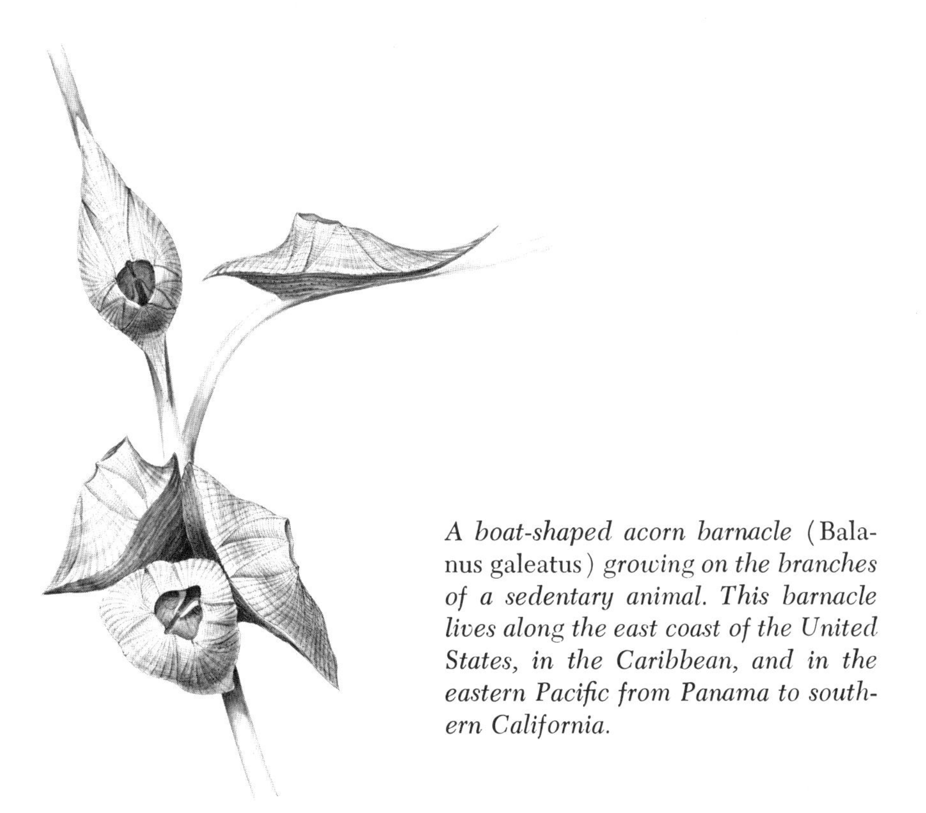

A boat-shaped acorn barnacle (Balanus galeatus) *growing on the branches of a sedentary animal. This barnacle lives along the east coast of the United States, in the Caribbean, and in the eastern Pacific from Panama to southern California.*

nineteenth century. This is not surprising, since their calcareous shells resemble some sea shells that occur in the intertidal zone—the transition area from the sea to the land that is alternately covered and uncovered once or, more commonly, twice daily by the tides. The true identity of barnacles was not established until their life history had been worked out in 1828 by a British physician, J. Vaughan Thompson, who discovered that barnacles were actually crustaceans masquerading in armorlike shells. We now classify the barnacles with the crustaceans, which includes, among other animals, the well-known crabs, shrimp, and lobsters. But in contrast to those crustaceans, barnacles do not molt their outer calcareous covering in order to grow larger. Instead, only the waterproof cuticle that covers the body and legs is periodically molted as the barnacles grow. Barnacle shells, like those of mollusks, are calcareous. They are secreted by the cuticle-covered animals within and enlarged by the addition of new layers along the edges.

It is not difficult to think of an adult barnacle as a shrimplike animal protected by limy plates. From the opening at the top of the shell an acorn barnacle can extend six pairs of leglike appendages, some of which serve to catch food, and some to pass it along to the mouth. The featherlike, curly legs inspired the use of the scientific name Cirripedia (*cirrus*, curl and *pedis*, of the foot) for the barnacles. As in the case of other crustaceans, the shell-less larvae swim freely for a time, but eventually they must settle on some hard object such as a rock. They then go through a complicated series of changes that leads to the shell-bearing adult. Once cemented to an object, they remain fixed to it for life. This highly specialized mode of living prompted the noted naturalist Louis Agassiz to describe aptly an acorn barnacle as "nothing more than a little shrimp-like animal, standing on its head in a limestone house and kicking food into its mouth."

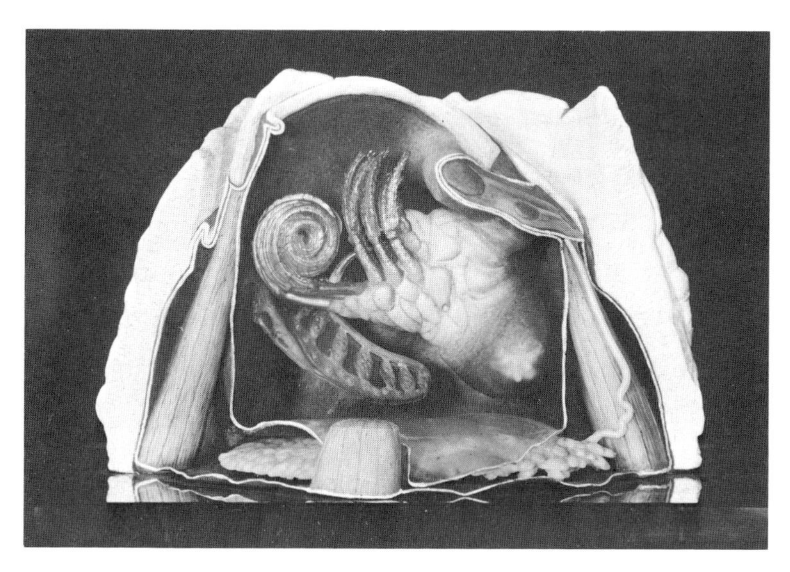

A museum model of a common Atlantic acorn barnacle (Balanus balanoides) *with one side of the shell removed to show the body and the cirri as they would appear when withdrawn into the shell.*

Barnacles have always fascinated mankind. Primitive man used them for food, and they are still eaten in certain parts of the world. In medieval times, the goose barnacle was thought to give birth to a bird, appropriately called the barnacle goose. Charles Darwin was so captivated by barnacles during the celebrated voyage of H.M.S. *Beagle* that after returning to England he spent eight years writing an exhaustive study of them. Scientists are presently trying to duplicate the cement secreted by barnacles. Barnacle adhesive can stick to almost anything, including glass, metal, concrete, and plastic, as well as natural objects. It is much stronger than any man-made adhesive. For this reason, it has been called "Superglue." Perhaps someday dentists may use barnacle cement to fix our teeth.

Barnacles have long been pests of the mariner. For more than two thousand years, man has sought to devise antifouling systems to keep barnacles from growing on his ocean-going

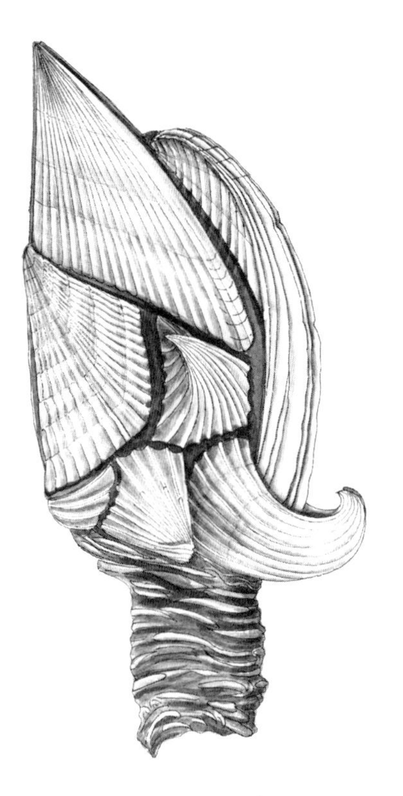

This well-armored, stalked barnacle (Arcoscalpellum multicostatum), *an inhabitant of the Antarctic Ocean, is about ½ inch tall.*

vessels. Large ships have been found to carry as much as three hundred tons of fouling organisms, mostly barnacles. Accumulations of barnacles on the hull reduce the ship's speed, increase fuel consumption, and cause frequent and expensive dry-docking. The final solution to marine fouling has still to be found. Scientists are continuing to study barnacles in the hope of learning better ways of controlling these pests.

After reading this brief introduction to the wonders of barnacles, we hope that you will want to learn more about these remarkable creatures. Perhaps on your next visit to the seashore you will make a special effort to observe these fascinating animals. If you are lucky, you may see one open its limy plates and reach out with its feathery legs in search of food!

2

Major Groups of Barnacles

The best known barnacles are the acorn barnacles, so named because of their superficial resemblance to the acorns that grow on oak trees. They are also known as rock barnacles, for they are commonly found on rocky shores around the world. Some of the lesser known barnacles in this group are the coral, sponge, turtle, and whale barnacles, which are named for the animals on which they may be found. Others are rarely obtained by collectors because they live far from shore, some at depths greater than seven thousand feet.

The acorn barnacle has a symmetrical, limy, outer covering. This calcareous shell is composed of a base and several neatly fitted plates that form an encircling wall to protect the soft body within. Shaped much like a miniature volcano, or tepee, the opening at the top leads into the body chamber, which is guarded by four limy, movable plates. These plates cover and protect the top of the animal, and they make up the operculum. In a similar manner, a limy or horny operculum covers and protects the aperture of many snail shells.

In contrast, the body of a goose barnacle is raised on a leathery or scaly stalk, and so it is called a stalked barnacle. The body may be covered with numerous calcareous plates, a few large plates, or just the vestiges of a few plates. The muscular stalk and the shell of the goose barnacle very much

An acorn barnacle (Balanus regalis) *from west Mexico, showing a top view of the shell (lower center, natural size), greatly enlarged views of the interior (top right and lower right), and exterior sides of the four opercular plates.*

resemble the gloved hand and forearm of the first baseman on a baseball team when he reaches out to catch a ball. The opening to the body chamber of this barnacle is on one edge, rather than at the top as in the acorn barnacle. Just as acorn barnacles are permanently cemented to some support, so too are the stalked barnacles. Some attach to ships, logs or other floating objects, and are carried to the far parts of the world. Most species of stalked barnacles, however, are found in the deep sea, some at depths as great as eighteen thousand feet.

The wart barnacles do not have a stalk, and they only superficially resemble acorn barnacles. The calcareous shell, rather than looking like a volcano, is more like an ornately carved box with a flip-top lid. Instead of four plates guarding the opening at the top that leads into the body chamber, only two make up the lid. Wart barnacles are rarely seen because they mostly occur in deep waters far from the shore. A few kinds live in the intertidal zone, but these are very small and hard to find.

Before we can discuss the classification of barnacles, we must establish their placement in the Animal Kingdom. Barnacles are

Left: *A side view of an inch-tall, deep sea stalked barnacle* (Calantica). *The shell of this barnacle, as with many stalked barnacles, is made up of numerous overlapping, limy plates. The stalk is well armored with plates also.* Right: *Wart barnacles* (Verruca) *growing on the spine of a sea urchin. Most wart barnacles are very small, generally less than 1/4 inch in diameter, and are commonly found only in deep water.*

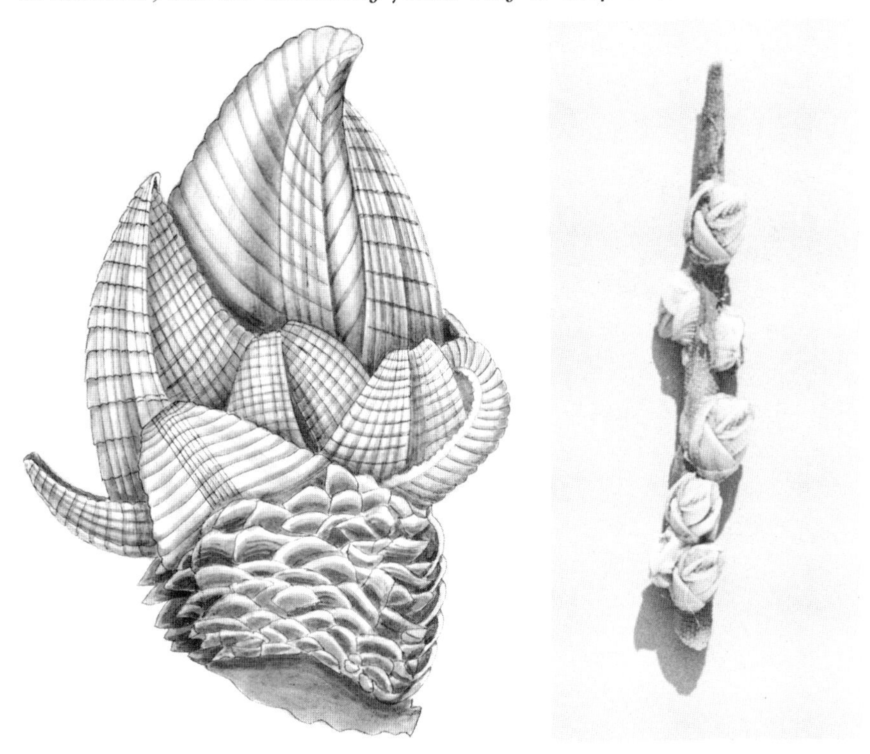

invertebrates, or animals without backbones. This is in contrast to the mammals, birds, reptiles, amphibians, and fish, which have an internal skeleton and are known as vertebrates. All barnacles and many other kinds of invertebrates have an outer covering known as an exoskeleton. It may serve as protective armor and it is especially noticeable in such animals as the lobsters, crabs, stony corals, and shelled mollusks and barnacles. Since most barnacles have a shell, which is part of the exoskeleton, early naturalists classified them with the mollusks. We now know that barnacles are related to the vast group of invertebrates which are placed in the phylum Arthropoda, the "joint-footed" animals. These include the insects, spiders, scorpions, centipedes, millipedes, crustaceans, and other less familiar animals.

Barnacles also belong to the superclass Crustacea, a name derived from the Latin word *crusta*, meaning a hard shell. Some of the more familiar crustaceans are the shrimp, crayfish, lobsters, crabs, and water fleas. Within the Crustacea there are several classes and the barnacles make up the class Cirripedia. This name refers to their curled feathery feet, called cirri, and barnacles are also known as cirripeds. Most of the barnacles we find along our shores have these unusual footlike appendages. Until recently, zoologists divided barnacles into five major groups or orders, but they now recognize only four. The presence or absence of a calcareous shell, as well as the differences in feeding habits, serve to separate these orders.

The acorn, wart, and stalked barnacles belong to the first of the four orders of barnacles. Because the segmented thoracic appendages or cirri are so characteristic of these barnacles, the group as a whole is known as the Thoracica. The cirri occur on each side of the thorax, and when they are extended from the shell they form a net that the barnacle uses to ensnare food. The cirri of the barnacle are precisely the same appendages that the crabs and lobsters use as walking legs. But in barnacles they have become modified for the unique purpose of providing a net to capture food. This change from walking legs to a feeding net evolved millions of years ago when barnacles gave up their free moving life for an attached or sessile existence. The thoracic barnacles are also distinguished from other groups of barnacles and nearly all other crustaceans by their calcareous shell, which they retain throughout their lives.

All the barnacles we have mentioned thus far, including the acorn, wart, goose, coral, sponge, turtle, and whale barnacles, are placed in the order Thoracica. The members of the second major group or order are unusually strange looking parasites that live on the "life blood" of other kinds of crustaceans, particularly on hermit and other crabs of one sort or another. These barnacles lack a calcareous shell or any vestige of one. They do not have cirri, nor do they have mouthparts with which to feed. These naked creatures obtain all of their nourishment from the host by a rootlike system which penetrates the host's tissue. As the root system was thought to have originated from the head

Left: *A ventral view of the blue crab* (Callinectes sapidus) *that has been parasitized by a shell-less rhizocephalan* (Sacculina). *We know that this parasite, which looks like a large, plump kidney bean, is a barnacle because it has larval stages similar to those of other barnacles.* Right: *The way to recognize the presence of the burrowing barnacles, acrothoracicans, is by the characteristically shaped openings of the burrows, such as those shown here, which are less than 1/8 inch long.*

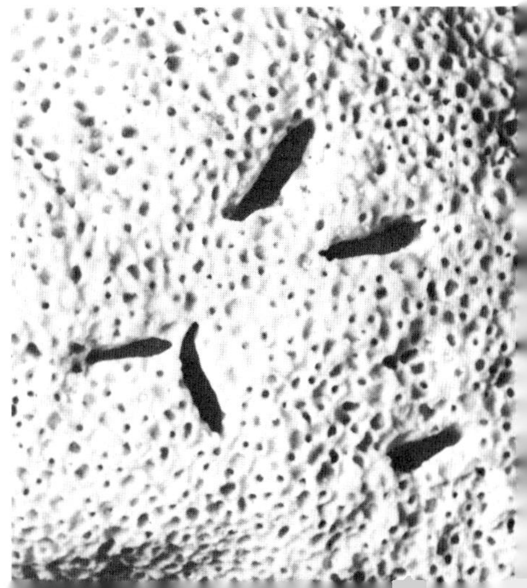

region, the name Rhizocephala or root-head, was applied to these barnacles. These parasites would hardly be recognized as barnacles, much less crustaceans, were it not for the fact that they have larval stages similar to those of other barnacles. Many rhizocephalans, especially those which become attached to the abdomen of the blue crab, resemble plump kidney beans. Others are shaped like long sausages or sacs.

The members of the third order of barnacles, the Acrothoracica, also lack an armament of calcareous plates. They make up for this lack by living in burrows that they excavate in limestone, stony corals, mollusk shells, and other invertebrates that have a calcareous outer covering, or exoskeleton. All of them have well-developed cirri and mouthparts for feeding in the fashion typical of the thoracicans. The acrothoracicans' name is derived from the Greek words *acro*, meaning extreme, and *thoracic*, meaning thorax or chest. This derivation refers to the position of the cirri which are bunched up at the extreme end of the chest.

The two dozen or so barnacles in the fourth and last order are parasites that lack a calcareous shell. Some live attached to the outside of the host animal and are called ectoparasites. Others live inside the host and are called endoparasites. Starfish, brittle stars, sea lilies, soft corals, and other similar animals serve as the hosts of these barnacles. The ectoparasitic kinds have only a few stunted appendages, which do not closely resemble those of the acorn or the stalked barnacles. And, because they live on the outside of their host, they have mouthparts which are used for piercing and sucking fluids from the body of the host. Those that live within the body cavity of their host, the endoparasites, obtain nourishment from the host's body fluids, which are absorbed by special lobes on the barnacle's body. These parasitic barnacles belong to the order Ascothoracica, meaning "baglike chest." They have either a

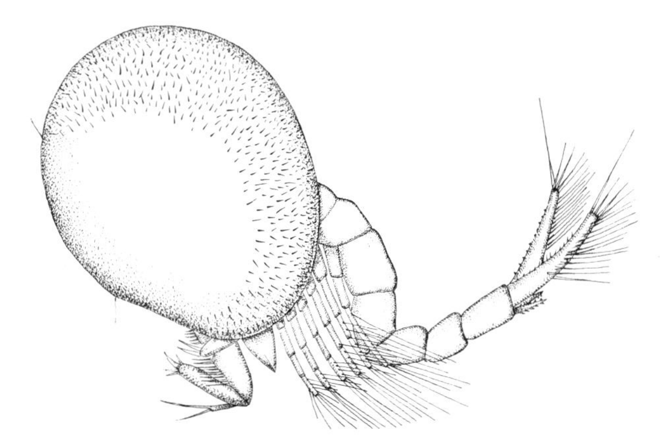

This ectoparasitic ascothoracican barnacle (Synagoga mira), *which is only 3/8 inch in diameter, lives on the tissue of soft corals.*

saclike body, or one enclosed by two valves resembling those of a clam. Unlike the clam's, however, these valves are not composed of calcareous material.

One of the remarkable features of parasitic barnacles and many other animals is that no matter how unusual or bizarre the adults may appear to us, the young or larval stages are always characteristic of the phylum to which they belong. Adult barnacles may not always be easily recognized. Careful study may be required to identify them correctly. This is especially true of the shell-less barnacles.

Now that we have seen how zoologists classify barnacles, let us describe in greater detail the various ways these remarkable creatures have become adapted, through the ages, to life in the seas.

3

The Life of Barnacles

Where They Live

Barnacles typically live in the sea. They have not adapted to life on land, or to continuous exposure to the air. Nor can they live in freshwater lakes and ponds, or in rivers and streams beyond the range of the inland flow of the tides. Some barnacles live in the mouths of rivers, where the sea is diluted by fresh water. Those that settle on the bodies of aquatic animals such as turtles may be transported far up rivers. Here they must either cope with a hostile, nearly fresh-water environment, or they die.

Wherever you go along the seashore you are almost certain to encounter some kind of a barnacle. This is especially true along rocky shores where barnacles may be found along the highest reaches of the intertidal zone. Barnacles have to be especially hardy to live in such a region, because they are exposed to the heat of the summer sun, as well as to cold winter rains and even blankets of snow. Yet barnacles have been able to adapt and survive under these conditions. What is even more remarkable is that some ten thousand barnacles may settle in an area of one square foot along a rocky shore, which doesn't leave room for many other kinds of animals.

Two billion barnacles in a narrow band a mile long poses many problems for future generations of barnacles and for many of the other inhabitants that are already living there. Overcrowding and competition for the meager amount of habitable space available mean a constant struggle for life. Similar conditions confront man, because he has created great problems for himself in his highly populated cities. Some acorn barnacles have evolved ways that aid in relieving crowding and in reducing competition for space. When acorn barnacles settle where there is ample space to grow, they usually have a volcano-shaped shell, wider at the bottom than at the top. Under crowded conditions, some individuals are able to grow normal looking shells by dominating their neighbors. This is done by either growing over them or by dislodging them from the places where they settle. In other cases, some barnacles have a small base and grow to look like tall slender cylinders. This mode of growth can be likened to building several skyscrapers

This coral-inhabiting acorn barnacle (Nobia grandis), *growing on a stony coral* (Euphyllia), *was collected on Warrior Reef, Torres Straits, Australia.*

on one small city lot. In yet other instances, a few crowded individuals will start out with a narrow base, grow tall and cylindrical for a while, and then expand the top part of their shells so that they flare out to look very much like the mouths of trumpets. Although barnacles are highly adaptable, they have not succeeded entirely in preventing overcrowding by subsequent generations. But then again neither has man.

The seashore is not the only place where barnacles live. Some kinds also occur in the depths of the seas, as well as on floating objects in the middle of the ocean. Not only do they live on immobile objects, but on fast-moving ships too. Barnacles have been known to settle on the pontoons of seaplanes, and then be transported to distant places. Under these circumstances, we can say that barnacles really take to the air.

In the sea, barnacles have taken up residence almost everywhere. Some live in sponges. Others are found on jellyfish, corals, starfish, sea biscuits, sand dollars, sea shells, worms, crabs, lobsters, horseshoe crabs, sharks and other fish, turtles, whales, and many other creatures. We should also remember that barnacles live in the polar regions of the world as well as in the tropical and temperate regions.

Some stalked barnacles build floats. The larvae of these barnacles settle on minute floating objects. As each barnacle grows in size, the antennae secrete a frothy looking, buoyant mass at the end of the stalk. This floating mass becomes a base for the attachment of other barnacles of its kind. It also provides a refuge for many different kinds of small animals. This barnacle is also known to settle on feathers floating in the ocean. Not surprisingly, a zoologist once described a barnacle under the name of *Ornitholepas*, meaning bird barnacle in Greek, when he found similar barnacles on the feathers of live seabirds. Another tiny stalked barnacle, *Koleolepas*, lives together with a sea anemone attached to the shell of a dead sea

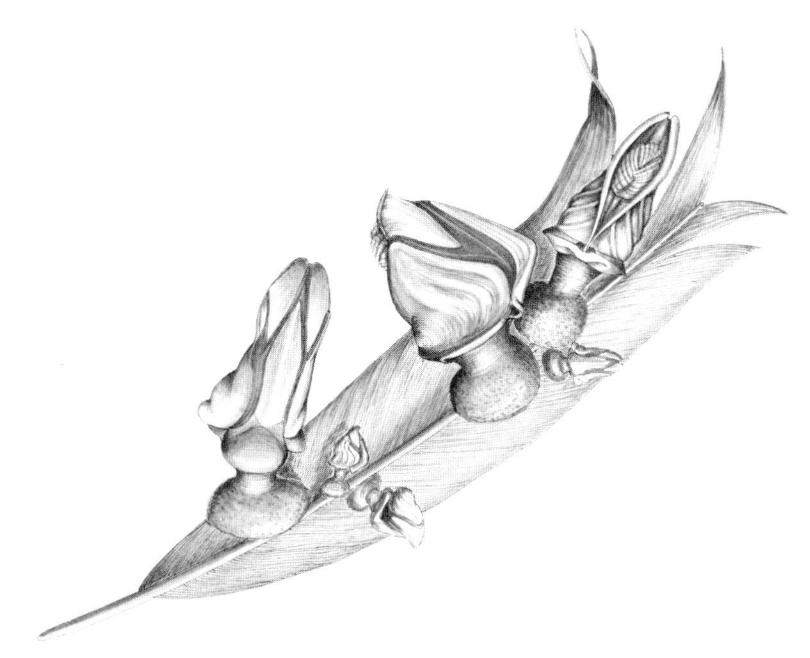

Different views of young and old individuals of a stalked barnacle (Lepas fascicularis) *found growing on a bird feather. The enlarged spongelike structure at the base of the stalk is a float secreted by the barnacle's antennae.*

snail. The snail shell serves as a house for a hermit crab, which carries its borrowed home, as well as the hitchhikers, wherever it goes.

How They Reproduce

Acorn barnacles are neither male, nor are they female, but rather they are both sexes. Zoologists call them monoecious, based on the Greek words meaning "one house," because each animal has both male and female reproductive organs. They also are called hermaphrodites, a word that is derived by combining the names of the Greek god Hermes and the goddess Aphrodite. In monoecious animals, each individual is capable

A clump of acorn barnacles (Balanus amphitrite hawaiiensis) *showing one with the sperm tube extended, and in the process of fertilizing a neighbor (top of picture).*

of producing offspring. Since each acorn barnacle has male and female organs, it might not seem unreasonable for an individual to fertilize itself. But this rarely happens, and when it does, it is only because another barnacle is not near enough to mate with it. Actually, we know that self-fertilization occurs in only two or three kinds of barnacles. As an interesting comparison, we should remember that most of the flowers in our gardens have both male and female organs in the same flower, but that cross-pollination is necessary for fertilization. As you are no doubt aware, this is where bees bearing pollen come into the picture. Of course, other insects also carry pollen, as does the wind. But in barnacles, what happens is that an individual fertilizes a neighbor and it is fertilized in return or by another neighbor. This is done by inserting an extendible tube carrying sperm into the body cavity of its neighbor. The sperm are then released from the tube, which may be an inch or more in length.

Many stalked barnacles are also monoecious, but in others the sexes are separate—that is, there are males and females. When the sexes are separate, the barnacles are called dioecious, a term derived from Greek words meaning "two houses," referring to the fact that each is of a different sex. In these barnacles, the males are much smaller than the females. In the animal world this notable size difference between sexes is not too unusual, and is commonly found in some deep-sea fish, many other crustaceans, and is of course frequently seen in spiders and insects. What is surprising is that the male barnacle's sole role in life is to fertilize the female and nothing more. This is because he is not endowed with any mouthparts or cirri with which to capture food. This means that he is also short-lived, for he cannot eat. As a result, the females live for several years and the males survive for but a few weeks or months at the most. Also remarkable is the fact that there may be more than one male associated with the female. In many cases, one female may have as many as six or eight males, and there are records of as many as one hundred males comprising the harem of the female.

In many of the stalked barnacles where the sexes are separate, the male lives in a small pouch in the wall of the body chamber near the female's aperture. Because the males of these barnacles are so small, Charles Darwin called them dwarf males.

In still other stalked barnacles, we find what are known as complemental males. This name was also coined by Charles Darwin. Complemental males are found in association with hermaphrodites, which, as you no doubt recall, have both sexes combined in each individual. So what then is the purpose of such males? Apparently they are not essential for the perpetuation of the race. But they probably serve the function of fertilizing the hermaphrodites when no other hermaphroditic individuals are available for mating.

In the acrothoracicans the sexes are separate, and, like the stalked barnacles, there are dwarf males. In the ascothoracicans and the rhizocephalans, the sexes are combined in some animals and separate in others.

The life of an acorn barnacle begins when the eggs are laid and fertilized by the sperm in the mantle cavity of the adult—in the space between the body and the bottom of the shell wall. The eggs grow in the mantle cavity where they are brooded until the time arrives for them to hatch. From fertilization to hatching takes about two months, but the period required for the embryo to develop varies greatly among species.

The newly hatched barnacle is called a nauplius. It is only 1/16 of an inch long and has a pair of large, hornlike structures, which serve to distinguish it from all other crustacean larvae. The minute first stage nauplius is a strange looking triangular creature that has three pairs of bristly appendages and a single median eye. Within minutes, or within a few hours at the very most, the nauplius sheds or molts its skinlike cuticle. The second stage nauplius is larger and the appendages are covered with even more bristles and spines. Four more naupliar stages follow, either hours or days apart, and each one is larger and more complex than the last. Between the third and fourth stages, the back develops a large protective shield or carapace that hangs out and over the sides of the body. In the middle of the carapace, there may be a long erect spine that points straight up or toward the back end of the body. By the time the sixth stage is reached, the nauplius develops a pair of large multi-faceted eyes, much like the compound eyes of flies. These are situated on each side of the median eye. The three pairs of appendages are used by the nauplius for a variety of functions. One pair serves largely for sensory purposes, and the others are utilized for swimming and for capturing food.

During the two or so weeks that the nauplius swims freely in the sea, it feeds voraciously on minute animals and plants that also swim or float about in the water. From the sixth stage on, there are two major changes in the life of a barnacle—the first

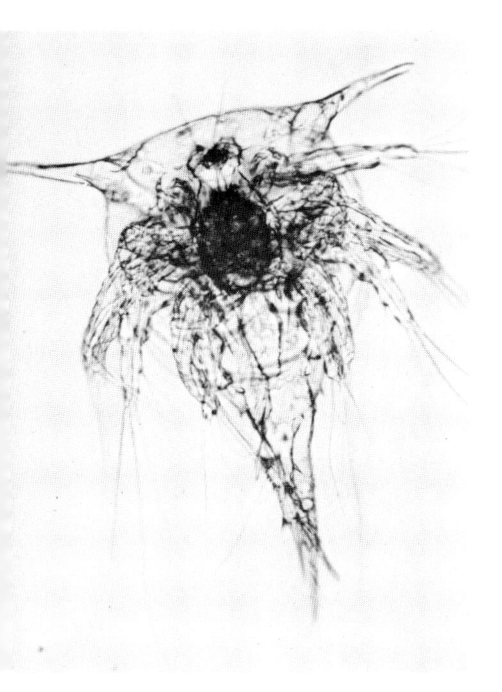

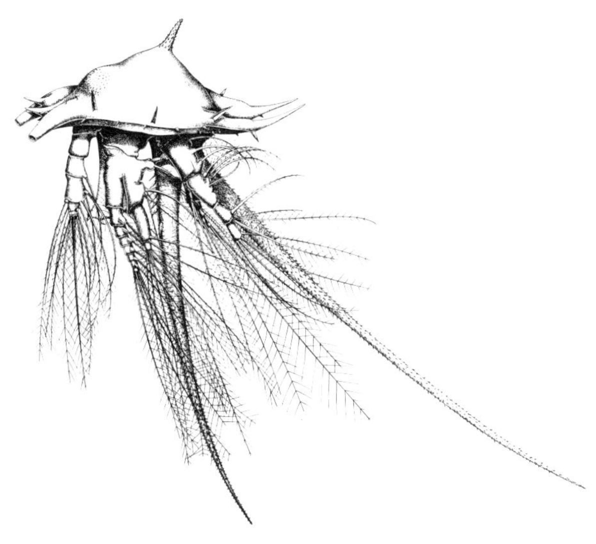

The last nauplius stage of a stalked barnacle, from the waters of the Antarctic Ocean. Natural size: 1/4 inch.

Left: *The last nauplius stage of an acorn barnacle showing the conspicuous lateral horns extending out from the sides at the top of the 1/8 inch long triangular body.*

is from a nauplius to a cypris, and the second from cypris to a pinhead-sized version of the adult.

After the sixth stage, the nauplius again molts, but this time it changes into a nonfeeding, weak-swimming animal that has a shell made of two valves which are hinged along the back, much like that of a clam. Not only does it have a bivalved shell, it now has three eyes and six pairs of legs. Oddly enough, at this stage of its life, the barnacle resembles a totally different kind of crustacean, an ostracod which is known by the scientific name *Cypris*. Because of the close resemblance to the ostracod, this stage of the barnacle is called a cypris. There is only one cypris stage. The main job of the cypris is to swim or crawl about in search of a place to settle down and start its adult life.

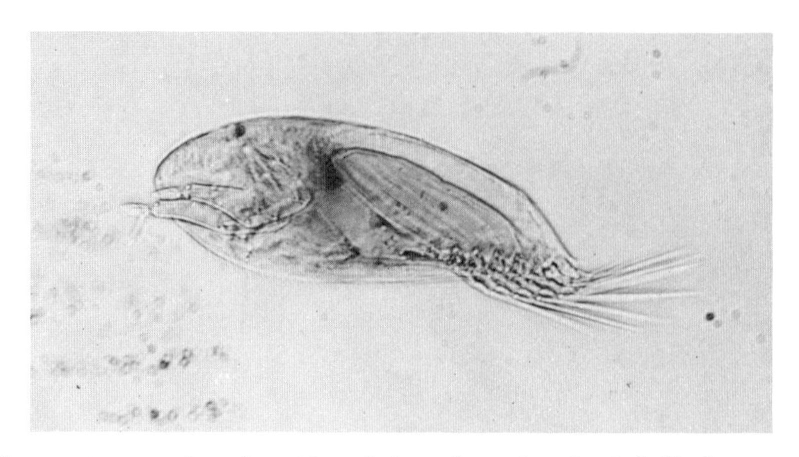

The cypris stage, less than ¼ inch long, has a bivalved shell, three eyes, well-developed antennae, which it uses to crawl about in search of a suitable place to settle down, and a full set of thoracic appendages.

The selection of a permanent building site for its home is an extremely critical stage in the life cycle of a barnacle. If it fails to find a suitable place to settle it will perish.

After searching for and finding a habitable site, often one close to others of its kind, the cypris secretes a little cement that anchors it in place for the rest of its life. The cement is secreted by glands that are located at the base of the appendagelike antennae. Once attached, the cypris undergoes the last of its major changes. These complicated changes involve a whole rearrangement of different parts of the body, all of which leads to the pinhead stage—a miniature version of the adult. Although this transition from the cypris to the adult is very much like the change we see when a caterpillar turns into a butterfly, there is a long period that the barnacle will have to go through before finally reaching adulthood.

The pinhead-sized barnacle grows rapidly. To make room for a larger body, the barnacle, like its relative the crab, molts at regular intervals. It sheds the thin, skinlike covering of the body, but not the limy, protective shell which it retains throughout life. The shell and the plates that make up the operculum are fashioned from the mantle which lines the inner surface of the shell and operculum. Naturally enough, the shell has to grow to accommodate the growing body. The calcareous shell is added to almost continuously by a secretion from the mantle, and the barnacle obtains this material from the water that surrounds it. In the case of the acorn barnacles and the wart barnacles, the plates which make up the shell grow upward from the base and along the margins of the individual plates so that the whole shell increases both in height and in diameter at the same time. In the stalked barnacles, new limy material is added to the individual plates on the underside, making them both thicker and wider.

From the time the egg hatches until the barnacle settles down and starts building its protective calcareous shell, three to five weeks have passed. Many barnacles live for only two or three years. But some may live as long as fifteen years, and in exceptional cases they may even reach an age of twenty-five years. By the time an acorn barnacle has reached such a ripe old age, it may have grown to be as much as four or five inches across, while some will attain nine or ten inches in height. As soon as an acorn barnacle reaches about one-half or slightly more of its adult size, it will begin to reproduce during the breeding season. And it will continue to breed each following season until it dies.

There are some exceptions to how soon and when a barnacle will reach breeding age. In warm waters, for example, the stalked barnacle *Lepas* may produce eggs four to five weeks after it has settled. These eggs may hatch after a week or so rather than after about two months. There is no set number of eggs that a barnacle will produce. Some species have as few as two dozen eggs at one time. Others may have as many as 110,000 or more. Some will produce a few hundred every few

A rare "little four plate" acorn barnacle (Tetraclitella divisa), *which is less than ½ inch in diameter, lives low in the intertidal zone in nearly all tropical seas.*

months throughout the year, especially those that live in tropical waters. The number of eggs that hatch and are shed into the sea as nauplii is astronomical, if we recall the number of barnacles that could occur on a one-mile stretch of rocky shore. But only a handful of these will survive the rigors of the sea to reach the settling stage and ultimately old age.

How They Feed

Once they become attached, barnacles cannot get up and move around in search of their food. Instead, they are dependent upon what the sea brings to their doorstep. Acorn barnacles have an ideal position for dining. They lie on their backs!

In the thoracicans and acrothoracicans, the appendages that occur on the thorax are generally flattened laterally and have a characteristically curled appearance. These appendages, commonly referred to as cirri, which, as we have seen, means curled, are composed of numerous segments, each one of which is covered with bristles or spines of varying lengths. When the

Two views of the red-striped (½ inch) California acorn barnacle (Balanus tintinnabulum californicus) *growing on a sea urchin (one inch). In the top photo the barnacle's opercular plates are closed, but in the bottom one they have started to open and the cirri are about to be extended.*

cirri are thrust out of the opening at the top of the shell in an acorn or a wart barnacle, or out of one edge of the shell in a stalked barnacle, they spread to form a netlike structure that sweeps through the water and entraps small animals and plants. The cirral net works very much like a fisherman's net, or like opening and spreading apart the fingers of your hand and then closing them, much as you would when you try to grasp something.

In *Balanus*, the common acorn barnacle, there are six cirri on each side of the thorax. This makes twelve altogether. But each cirrus has two arms, or rami, making a total of twenty-four arms in all. The cirri closest to the mouth are the shortest, and proceeding back along the thorax, the cirri become progressively longer.

In some barnacles the cirri closest to the mouth do not sweep the water. These cirri are used to clean off the prey passed down by the other cirri. In the acorn barnacles *Tetraclita* and *Chthamalus*, there are little spines that have been modified so that they look like small combs. These are highly efficient in cleaning off the food entwined and stuck on the bristles and spines that clothe the cirri making up the feeding net. The short cirri then transfer the food to the oral appendages, which in turn tear it up and stuff it into the mouth. The food consists largely of copepods, diatoms, the larval stages of crabs and other crustaceans, and even the larvae of barnacles—almost any organism small enough to be entrapped and engulfed. The next time you are at the shore, wave your hand between the sun and a feeding barnacle. If a shadow is cast on the barnacle while feeding, then all of the feeding activities will cease and the net will be withdrawn. This defensive response is known as a shadow reflex.

In most cases, the cirral net of the acorn barnacle can be rotated either to the left or to the right through 180 degrees. This allows the barnacle to take advantage of the flow and ebb of the sea as it sweeps past and eddies around the shell. In the

Two views of a museum model of a common Atlantic acorn barnacle (Balanus balanoides).

stalked barnacle there is no need for the cirral net to swivel around, because the stalk is capable of doing this job. If the water is not flowing rapidly, the barnacle makes repeated sweeps —forty to fifty times or more each minute—until the net has become sufficiently filled and is ready to be emptied. If the water passing the net is moving rapidly, the barnacle may just extend the net and allow the current to fill it without bothering to work too hard. *Lepas*, a stalked barnacle, is more aggressive than the acorn barnacle in feeding habits. Because it occurs on floating objects moving with the currents, its flexible stalk must swing in all directions while the cirri grope for food. The cirri may work in unison, or independently of one another. When food is abundant, individual cirri may grasp and hold prey until it can be conveyed to the mouth.

Barnacles exposed by the receding tide cannot feed until the tide returns and they are covered by the sea. Those that live in the intertidal zone are restricted to eating when covered by water, during the high tides. This is an enforced method of dieting. Barnacles permanently housed below the surf zone, however, are not so restricted in their dining hours, and they may feed whenever the need arises.

There is one kind of acorn barnacle that lives on a coral which is found only in a few places in the Indian Ocean. It is a rather unique creature because it does not have cirri with which to gather food. Instead it has evolved a remarkable method of feeding that insures there will always be food present when it comes time to eat. This barnacle, known as *Hoekia*, in honor of the Dutch zoologist Paulus Peronius Cato Hoek, has developed a way to harvest the soft body tissue of corals. We do not know all the details, but we do know that the coral's tissues grow over

A coral head from a reef in the tropical waters off Florida, with "horn shell" coral barnacles (Ceratoconcha) *partly overgrown by the coral. The barnacles have diamond-shaped or oval apertures.*

the opening of the barnacle and cover its opercular plates. When the barnacle needs to feed, it opens the operculum and extends its mouthparts. These are flat and long, with one pair looking very much like the blade of a cross-cut saw and the other pair looking like the blade of a ripsaw. Whenever these sawlike mouthparts come in contact with the soft tissue of the coral, they saw off a hunk, enough for a meal. While the barnacle is eating and digesting its meal, the coral sets to work to repair the damage. The coral eventually grows new tissue, only to have the newly regenerated soft parts eaten by the barnacle.

Not all barnacles have to work for their dinner. Indeed, some do not even have to wait for currents to stir up enough food for them to capture. The whale barnacles, such as *Coronula*, merely have to open the door and put out their cirri. The whales do all the work, and the barnacles merely sit back and wait for the currents created by the whale as it moves to bring them food. In the rabbit-eared whale barnacle, *Conchoderma*, the water flows through the cirral net so fast that it has evolved a method to keep the water from tearing it loose from its mooring. After the water passes through the feeding net, it enters the body chamber and is then funneled out of the openings at the top of its earlike lobes.

The parasitic rhizocephalans do not have cirri or mouthparts with which to feed. Instead they have a root system, much like that of a tree, which invades the tissues of the host. Their food is derived from the tissues of the host. The result is that these barnacles always have a supply of food as long as the host lives, but the host must constantly forage for its own meals.

Their Enemies and How They Defend Themselves

The barnacles found along rocky shores, on wharf pilings, and elsewhere would seem to be easy prey for any hungry animal looking for a meal. But this is not true, because most barnacles have evolved some clever ways to protect themselves. They are not always successful, but they have managed to survive the onslaught of predatory animals that began when barnacles first appeared in the seas, some 400 million years ago. Among the many predators of barnacles are flatworms, nudibranchs and other gastropod mollusks, annelid worms, centipedes, crabs, starfish, sea urchins, fish, and many birds, such as the oyster-catchers, turnstones, surfbirds, and gulls.

The primary defense of most kinds of acorn barnacles is to quickly draw in their cirri and close their opercular plates. Once shut up inside their protective shell, they are sealed off from the outside world. But how do they defend themselves from the attack of boring predators such as the gastropod mollusk? This problem is solved by some barnacles that have either a very thick shell or a shell composed of layers of calcium carbonate and a tough elastic tissue called chitin. Borers soon give up trying to penetrate thick-shelled barnacles, because they expend too much energy trying to drill through the shell. Sometimes they spend hours drilling through a shell only to find that there is no barnacle tissue within. Some other predator had been there first, or the barnacle had died of natural causes. Some acorn barnacles have a laminated shell of calcite and chitin, which is like a sheet of plywood, with the wood representing the calcite and the man-made glue holding the layers together representing the chitin. This shell deters some borers from penetrating too deeply. It is probably the chitin that impedes the borers, perhaps because it clogs up their filelike teeth.

Other predaceous snails may crawl up the shell of the barnacle and perch over the opercular plates. These gastropods pull the opercular plates apart by exerting pressure with their foot, or by inserting a rodlike spike into the aperture of the barnacle's shell. Once there is enough space between the opercular plates, the snail inserts its tongue, which is covered with row after row

Left: *The predaceous zebra thorn snail* (Opeatostoma pseudodon), *which lives in the eastern Pacific from west Mexico to Peru, uses its ½-inch long, daggerlike apertural tooth to penetrate and pull apart the opercular plates of barnacles.* Right: *An acorn barnacle* (Balanus peninsularis), *from Cape San Lucas, Baja California, Mexico, has protective spines that serve to deter predators from crawling up the sides of the shell.*

of tiny sharp teeth. This highly efficient organ, commonly known as the radula, moves back and forth, and works much like a wood rasp or file.

The predatory starfish will also climb onto the shell of an acorn barnacle and try to gain entry through the operculum. It uses several of its tube feet to pull apart the opercular plates. And, like the gastropod, it slowly and steadily exerts pressure until the barnacle can no longer keep the operculum closed. But unlike the gastropod, which uses its radular teeth to rasp away the tissue, the starfish pushes its stomach into the shell and slowly digests the barnacle's soft body. Some starfish may eat as many as fifty small barnacles a day.

At least one barnacle has evolved a method to keep animals from crawling up its shell. It has downward projecting spines on the shell that deter crawling predators.

Many invertebrates and passing fish will nip at the extended cirri of barnacles in an effort to obtain a snack to tide them over until a better meal can be found. The cirri eventually grow back, but for a time the barnacle's ability to catch food is seriously impaired.

Crabs that eat barnacles just crush the shell in their strong pinchers and then eat them, discarding the inedible shell as they consume their meal. There is really no defense against the crushing claws of a crab, and luckily for the barnacles there are not many crabs that relish them.

Some time in the past, acorn barnacles evolved a tricky method of getting rid of small animals that creep or crawl onto their opercular plates. This is what they do. When an annelid worm, for example, comes in contact with the opercular plates while looking for a place to rest or hide, the barnacle responds by rotating the opercular plates against the wall of its shell. As a result, the unwanted intruder is ground up, if it is not quick enough to escape. The barnacle's ability to grind up unwanted intruders enables it to keep the opercular region clear, and it is thus able to extend its cirri and fish for food.

Rock grazing or browsing animals such as snails, sea urchins, and many fish are also major predators of young barnacles. They don't set out especially to make a meal of barnacles, but being rather unselective about what they eat, they may gobble up barnacles, especially those that are newly settled. One browsing fish known as the sheepshead or convict fish, because it has a number of dark stripes on its side, has strong unnotched incisor teeth which are used to scrape barnacles off the rocks. The parrot fish of tropical waters also feeds on barnacles. Random feeders, such as these fish, can clean off large areas of rock and decimate local barnacle populations.

The best defense at times is having no defense at all. This is especially the case for those barnacles that live on or in other

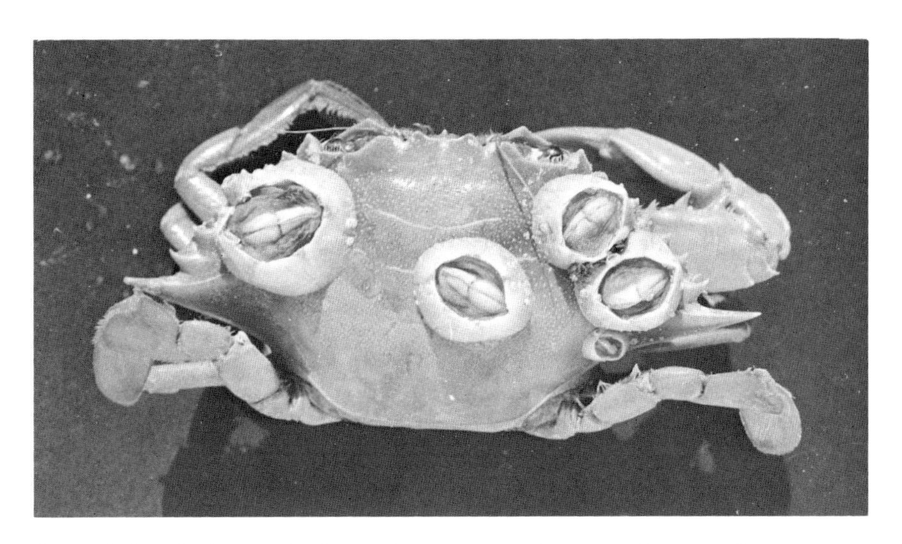

Several "turtle barnacles" (Chelonibia patula) *growing on the carapace of the blue crab* (Callinectes sapidus), *collected in shallow water on the Gulf coast of Florida.*

animals. The root-headed rhizocephalans certainly do not need any defensive armament since their host protects them. This is especially true of those host-crabs that have large pinchers. The "turtle" barnacle, *Chelonibia patula*, which actually attaches to crabs, also has no need for any special protection since its host takes care of this job. Any threat or potential danger to the crab is also a danger to the rhizocephalan and to the turtle barnacle. Those stalked barnacles that attach to the gills of crabs and lobsters are similarly protected by their hosts, as no doubt are the stalked barnacles that settle on jellyfish, which have highly poisonous stinging cells to defend themselves. In much the same way, whale barnacles have no need for any special defenses. After all, what animal is large enough to take on a whale? Even the giant squids, which may attain sixty feet in length, cannot compete with this mammoth of the ocean. They generally end up in the stomach of the whale after a battle in the depths of the sea.

4

Myths and Tales

The Barnacle Goose of British Mythology

The term goose barnacle dates back to the folklore and mythology of the sixteenth century and even earlier. The myth, as we shall soon relate, was accepted by many European naturalists of the time. They undoubtedly saw the remains of tree trunks washed ashore which had been afloat in the sea for many months. They found attached to these floating logs stalked barnacles which were mistaken for the fruit of the tree. These early naturalists also believed that the cirri protruding from the goose barnacle's body looked like the tail feathers of a young bird, and that the long stalk of the barnacle actually represented the neck of a goose. The result of this absurdity was simply their belief that a seabird called the barnacle goose came from the fruit of a tree that grew along the shores of the sea. The myth may also have been invented to account for the fact that some geese, being migratory, were never seen to breed in England, and many barnacles do have the general shape and coloration of wild geese with their wings folded.

The myth appears in many books written in the 1500s and 1600s. With the rapid progress being made in the art of printing, the myth spread like wildfire during this period. The best rendition is found in *The Herball or General Historie of Plantes* of John Gerard, which was published in London, in 1597. One

Goose barnacles (Lepas fascicularis), *many with their cirri extended, are shown growing on kelp. This sight prompted early naturalists to confuse the cirri with the tail feathers of a baby bird.*

chapter is entitled "Of the Goosetree, Barnackle tree, or the tree bearing Geese." From it we quote, in the original English spelling, the following account: "There are found in the North parts of Scotland and the Islands adjacent, called Orchades, certaine trees whereon do grow certaine shells of a white color tending to russet, wherein are contained little living creatures: which shells in time of maturity doe open, and out of them grow those living things which, falling into the water, doe become fowles, which we call Barnacles: in the North of England, brant geese; and in Lancashire tree geese; but the others that doe fall upon the land perish and come to nothing."

The myth was by no means limited to England. Variations and many elaborations appeared in the writings of the time and for more than 150 years after it appeared in the herbal of Gerard.

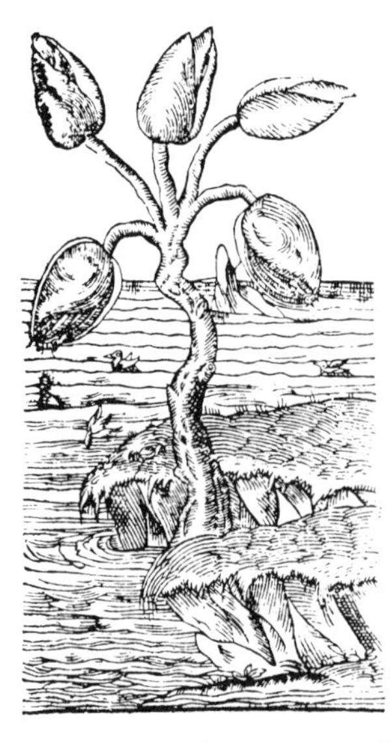

The barnacle tree from the 1597 edition of the Herball *of John Gerard, with "fruit" which resembles the stalked goose barnacle that we now call* Lepas.

A natural outgrowth of the barnacle goose myth was the belief that the goose does not "tread [walk] as other fowles do, they lat [laid] no eggs, they never sitte [on the eggs to hatch them]. And therefore in certain places of Ireland in Lent and upon other fasting daies [days], it is lawfull to eat these fowles, as if they were not flesh, nor came of flesh." As late as the early 1900s, it was still considered permissible to eat the succulent barnacle goose during the Lenten season as an ancient and established custom.

Darwin's Animals

As we mentioned earlier, it was not until 1828 that the British physician J. Vaughan Thompson worked out the complicated life history of barnacles. He demonstrated that their young stages were similar to those of other crustaceans, and that barnacles were not mollusks as was previously believed. This established their systematic position as a highly specialized group of crustaceans. Thompson's findings were published in 1829, just two years before Charles Darwin boarded H.M.S. *Beagle* on a voyage that would take him around the world. The *Beagle* was to spend five years away from England, and Darwin, who was a few months short of being twenty-three years old, was to accompany the ship as an unpaid naturalist, something unheard of today. Darwin first became attracted to barnacles when the *Beagle* visited the coast of Chile in 1834. It was here that he found a "most curious form" burrowing in the shells of snails. Thereafter, he made a special effort to observe and collect barnacles wherever he encountered them during the cruise.

The *Beagle* returned to England in 1836, but it was not until ten years later that Darwin began his studies on the barnacles, and he was to devote almost eight years to the task of monographing the barnacles of the world. By the time he was ready to work on the barnacles, he had borrowed or begged specimens from almost every naturalist that he could, and he claimed that he had amassed a collection of over ten thousand specimens. Much of Darwin's house was devoted to the storage and study of his "beloved barnacles." There is a story that describes the influence this work had on Darwin's family. One of his children was invited by a young friend to tea. Darwin's child asked about the friend's father, especially asking, "Where does he do his barnacles?" The child assumed that every father studied barnacles.

Between 1851 and 1854, Darwin produced four volumes describing all the known living and fossil species of barnacles. These monographs formed the basis for the modern classification of these crustaceans, and they are still considered the standard works of reference on the subject. The many years that Darwin devoted to the barnacles undoubtedly proved invaluable when it came to formulating the "species concept" that eventually lead to his theory of natural selection. Thus, barnacles have come to be known as Darwin's Animals. Perhaps, if Charles Darwin had not been inspired to study these crustaceans, he would not have achieved the scientific recognition of his colleagues, and his ideas and theory on evolution might have been largely ignored by the scientific world.

5

Some Curious Barnacles

Hitchhikers of the Sea

Barnacles are some of the most tenacious hitchhikers found in the sea. They settle on the bodies of a variety of strange animals, including whales, manatees, dugongs, turtles, sea snakes, and fish. These barnacles live commensally with their host. That is, they receive some benefit from their host, such as sharing food, without harming or interfering with the life processes of the host. Others are parasites that live at the expense of their host. In both instances, these barnacles have evolved highly specialized modes of life and have become dependent on the host animal. Without the host they cannot survive, and they perish when the host dies.

Whale Barnacles

Whales are the largest creatures in the world today. Blue whales are the goliaths of the sea. These mammals can exceed 100 feet in length and weigh more than 150 tons. Regardless of their size, the blue whale and several other kinds of whales carry around different kinds of hitchhikers. One of these is an oval-shaped, flat-bodied, isopod crustacean known as the whale louse. Of interest to us are the hitchhiking acorn barnacles. These are the most unwanted hitchhikers because they grow into the whale's skin. Once they become firmly anchored, they are not easily dislodged, even when the whale swims through the water.

Barnacles do not grow just anywhere on a whale. They are found on certain areas of the skin that will allow their cirral net to be most directly exposed to food-laden currents. They also occur only on body surfaces where they do not inhibit the whale's movement in the water, such as on the head, flippers, lower lip, between the eye and the base of the flippers, and on the tail region.

One question has puzzled zoologists for many years: When do the barnacles settle on the whales? In December, January, and early February of every year, more than eight thousand California gray whales, for example, congregate in the shallow bays, lagoons, and near shore waters along the Pacific coast of Baja

The noted biologist, Dr. Carl L. Hubbs, examining a gray whale that was stranded on the beach near San Diego, California. The head region of the whale is heavily encrusted with barnacles.

California, in Mexico. They leave the Bering Sea off Alaska in the fall and migrate southward on a two to three month trip of some five to seven thousand miles to breed and to give birth to their young. Perhaps the answer to the question requires us to recall that the larval barnacles are released into the sea, where they swim and float and eventually reach a stage when they are ready to settle permanently on a suitable substratum. Therefore, it is not difficult to envision that the larvae are released into the sea at the same time when large herds of whales congregate at the breeding grounds. The twelve- to seventeen-foot-long baby whales, or calves as they are called, are first exposed to the larval barnacles at this time. If they should manage not to acquire any barnacles shortly after they are born, then they are likely to be infested the following year when the herd returns to its winter

A California gray whale inhaling before diving below the surface off the coast of San Diego, California. The irregular, whitish patches on the head and "shoulders" (left) are whale barnacles.

breeding grounds. Thus it appears that the breeding seasons of these barnacles are synchronous with those of the whales.

Some whales also carry colonies of stalked barnacles as well as acorn barnacles. One of these stalked barnacles bears the name rabbit-eared barnacle, because it has protruding from the top of the body two large flaps of what look much like the ears of a jackrabbit. This barnacle generally does not settle directly on the skin of the whale, but instead settles on the shell of an acorn barnacle which has attached to the whale. There may be as many as six, eight, or possibly more rabbit-eared barnacles attached to one acorn barnacle. As the whale moves through the water, the cirri of the rabbit-eared barnacle extend out a short distance to capture food. When water flows through the cirral net, the food particles are retained, and the water exits through the holes at the tips of the earlike projections. If no holes existed, the force of the water might tear the barnacle loose from its support. This is a good example of how an animal has adapted to a highly specialized way of life.

Turtle Barnacles

Sea turtles have a long geologic history that dates back more than 100 million years. Some time in the distant past, barnacles became hitchhikers on these marine reptiles. Such barnacles are most appropriately called turtle barnacles, although not all species of turtle barnacles actually live on turtles. Some, as we will see, settle on sea snakes and crabs because these creatures provide a habitat similar to that of turtles.

Turtle barnacles differ from whale barnacles only in the manner they use to attach to their host, and in certain details of the way their shells grow. But, like the whale barnacles, they must settle on areas of the turtle which will permit the capture of food with their cirral net as the turtle swims through the water. Not all turtle barnacles are found on the turtle's shell. Some kinds

A whale barnacle, the diadem crown (Coronula diadema), *removed from the skin of a humpback whale captured on the coast of California. The strange looking, five- to six-inch-tall stalked barnacles growing on the diadem crown are rabbit-eared barnacles* (Conchoderma auritum).

settle on the leathery skin of the flippers and head, while others even settle in the mouth and throat of the turtle.

Turtle barnacles, like the whale barnacles, apparently breed only at the same time of the year as their respective hosts do. But there is one major difference. The young turtles are not infested by barnacles at this time. This is because the barnacle-encrusted female, after she struggles ashore to lay her eggs in the sand high up on the beach, returns to the sea and then swims off to a distant feeding ground. Incubated in the warm sand, the eggs hatch some two months later. The baby turtles, which are only two or three inches long, dig their way out of the nest buried beneath the sand and move toward the sea. Many never reach the sea, for ghost crabs, birds, dogs, and other animals prey on the young turtles. After safely reaching the sea, many are eaten by fish and other predatory animals. Those that manage to survive swim off in search of the feeding grounds. They do not return to the breeding grounds of their parents until they reach sexual maturity, some three or four years later. At this time they are first exposed to the larvae released from the barnacles living on the older turtles, and the young turtles then acquire their hitchhiking barnacles.

One of the most bizarre hosts is the sea snake, which is also a marine reptile. One kind of "turtle" barnacle has adapted to living on the yellow-bellied sea snake, which inhabits the central Pacific Ocean. Some are so heavily encrusted by barnacles that one wonders how they are able to swim. In the New World, sea snakes occur only in tropical Pacific waters as far north as Mexico. Care must be taken when handling these reptiles, for they are highly venomous, even more so than rattlesnakes.

Parasitic Barnacles

We have previously discussed the various kinds of parasitic barnacles. One of these shell-less barnacles, a rhizocephalan, has also evolved a curious hitchhiking habit. It has taken up residence as a parasite of a crab that lives in many of the rivers and streams along the coast of Japan. In order to spawn, the host crab must migrate once each year to the more saline waters of the sea. Like other hitchhiking barnacles, the breeding periods of this parasite and its host occur at the same time. As a result, the barnacle is able to infest other crabs during their spawning season. These barnacles are dependent upon the crabs to transport them annually to the sea where they both breed. This is nature's way of perpetuating these hitchhikers.

6

Useful Barnacles

Barnacles are often looked upon as being of little value to us in our everyday needs. But we must remember that barnacles are important in the overall economy of the sea, because they form a link in the intricate web of life. They harvest small plants and animals for their food, and in turn they serve as food for other animals. Furthermore, the larval stages of barnacles form an important part of the food chains that support still other marine organisms, such as sea anemones, corals, sea squirts, larval fish, and so forth. The limy skeletons of adult barnacles also provide a base for the attachment of these and many other kinds of animals as well as plants.

Barnacles We Eat

In many places around the world, barnacles are eaten much as we eat shrimp, crabs, lobsters, and other forms of shellfish. The barnacles that are eaten are selected either for their large size or their great abundance.

In the Pacific Northwest, the American Indians used to eat an acorn barnacle, *Balanus nubilus*, which reaches a gigantic size. Remains of this barnacle occur in the kitchen middens of the ancient campsites of the Indians. In Chile and Peru, people still make a stew from another acorn barnacle, *Balanus psittacus.*

This barnacle attains a height of six or seven inches and a diameter of two or three inches. Some individuals grow even larger. Because most of the barnacle is inedible shell, many individuals are needed to make a meal. In many markets in Chile and Peru you can buy fresh barnacles, or you can buy them already canned and ready to eat.

The stalked barnacles, including the commonly found *Pollicipes* and *Lepas*, are excellent to eat, and they taste very much like lobsters. Instead of the body, only the large meaty stalk is consumed. After you have collected as many as you want to eat, here is how to prepare them: Rinse the barnacles in fresh water and let them drain. Then drop them in a pot of boiling water seasoned with a little salt and a bay leaf. Boil them for three to five minutes and then remove the pot from the heat. Drain off the water and let them cool sufficiently before peeling off the tough outer covering of the stalk, just as you would peel the shell off a shrimp. Discard the body and its covering of shelly plates. The meaty part of the stalk is now ready to eat.

In San Francisco, around the turn of the century, it was the practice of the local chefs to take the boiled barnacles and season them with butter, a little fresh parsley, and a pinch of garlic, and then steam them for a few minutes, adding lime or lemon juice before serving. Cooked barnacles can also be eaten chilled like shrimp, and dipped in a mixture of ketchup, chili sauce, and horseradish, or they can be eaten warm like boiled lobster and dipped in melted butter. This exotic cuisine is an epicurean delight according to some gourmets, but it is not likely to replace meat and potatoes in our diet.

Barnacle Cement, Nature's Superglue

Anyone who has tried to pry an acorn barnacle from a rock knows how firmly they are attached. When the cyprid larva finally finds a suitable place to settle, it secretes a cement that permanently attaches it to the spot. What is remarkable about this cement is that it can anchor a barnacle for a long period of time, even after the barnacle dies. Furthermore, it can anchor a barnacle to almost any hard substance. The list of objects on which a barnacle can settle is quite extensive, and includes such man-made things as light bulbs, bottles, phonograph records, non-stick frying pans, and, of course, natural objects such as rocks, crabs, whales—even the feet of penguins. The list is almost endless.

The tenacity of barnacles once affixed to an object has long been of interest to scientists. Man has been unsuccessful in his attempts to duplicate the adhesive or to find ways to dissolve it. Barnacle cement hardens like many plastics, even under water, and has certain physical properties that dental and medical researchers have long sought after, namely great strength, and resistance to temperature variations and to solubility. The cement is twice as strong as the cements used to hold our spaceships together. It is four times stronger than the glue used in making furniture. Barnacle cement heated to 622° Fahrenheit merely softens; it does not melt. At the other end of the scale, it does not peel or crack when chilled to 383° Fahrenheit below zero. Tests that have been made to dissolve the cement have also proven to be failures.

Much remains to be learned about barnacle "Superglue" and its use in medicine and dentistry. We do know that the adhesive secreted by barnacles has a very complex molecular composition, and consists largely of protein. Should we find a way to duplicate it, such a glue could be used in medical surgery for repairing bones and especially in dentistry. Scientists believe the moisture ever present in the mouth would increase the adhesive property of this cement. Remember, barnacle cement is impervious to fluids. It is also resistant to the bacteria that cause teeth to decay!

Barnacles in Medical Research

Because barnacles are easy to obtain and because many of them grow to a large size, their use in medical research has become very popular. Many scientists are studying the chemical and structural properties of muscle fibers with the desire of applying the results of their work to solving some of the diseases that afflict man. The giant barnacle, *Balanus nubilus*, has muscle fibers of immense size. They rank among the largest muscle fibers found in the animal world, ranging from fourteen to twenty times thicker than those of humans. Biologists use them to conduct experiments on the nature of muscle contraction. By injecting living barnacle fibers with chemical tracing agents and then by stimulating the fibers with electrical charges, scientists are able to understand better the events that cause muscles to contract.

Barnacle Fertilizer

Barnacles are used as fertilizer to help farmers grow better crops. In Japan, bundles of bamboo sticks are anchored on the floor of the sea near the coast. After a while, barnacles settle on the bamboo, where they are left to grow until large enough to be harvested. This may require a year or two. The bamboo bundles are then taken from the sea, the barnacles are removed, broken, and dried in the sun. This mineral-rich product is used by farmers as a source of fertilizer.

7

Harmful Barnacles

Boring Barnacles

Geologists and engineers have long recognized that rock-boring plants and animals are responsible for much of the erosion of rocks along coastal areas. Barnacles, however, are seldom thought of as organisms that cause erosion by boring into rocky shorelines. Nevertheless, one of the most interesting and important boring animals is the stalked barnacle *Lithotrya*, which is found in shallow waters and in the intertidal zone of tropical seas. The scientific name of this barnacle is very appropriately derived from the Greek words *lithos*, meaning stone and *tryo*, meaning to rub.

Lithotrya makes its home in limestone rocks by mechanically wearing away the rock. The barnacle excavates its burrow by rubbing against the rock with hundreds of minute calcareous plates that arm its stalk. Initially the burrow is small, but as the barnacle grows and needs a bigger home it enlarges the burrow. The burrow of an adult is about ½ inch wide and penetrates three inches deep into the surface of the rocks. As many as two or three burrows and even more may cover each square inch of limestone. Because *Lithotrya* live so close together, the burrows riddle the rock, which eventually becomes honey-combed, in much the same manner as termites tunnel out wood beams in a house.

A piece of limestone from the Bahama Islands with several burrows made by the rock-boring stalked barnacle (Lithotrya). *One of these two-inch-long barnacles remains in a burrow.*

A colony of *Lithotrya* will excavate about one third or more of the limestone rock in which they make their burrows. Eventually these burrows weaken the rock which may crumble and become part of the bottom sediments. In this way, the barnacles join forces with the pounding surf and scouring wind to erode away rocky shores.

The acrothoracicans are also boring barnacles, but they have a preference for boring into the limy shells or skeletons of other organisms, notably corals, mollusks, and echinoderms. The acrothoracicans do not have a rasping stalk like *Lithotrya*, but they do have certain areas of their body covered with small, sharp, toothlike spines that are used in excavating burrows. The burrows of these shell-less barnacles serve to protect their naked bodies in the absence of an armorlike exoskeleton. Because acrothoracicans rarely grow larger than 1/8 of an inch, a single gastropod shell, for example, may provide homes for many dozens of individual barnacles.

Normally, acrothoracicans do little if any harm to their host, but on the other hand, they contribute nothing to their host's well being. Severe infestations of these barnacles may structurally weaken the host's shell. If they burrow into a clam, it may secrete blister pearls to cover the burrows, but only if these penetrate the inner surface of the clam's valves.

Ship Fouling

The settlement and growth of barnacles and other marine organisms on the surface of submerged man-made objects is known as fouling. In tropical waters the fouling of a ship's hull may take only a few months, but in colder waters it may take six months or up to two years. Acorn barnacles undoubtedly are the most troublesome of the fouling pests. Goose barnacles may present a problem for wooden-hulled sailing vessels, but they are not commonly found on fast-moving motorized ships.

When an ocean-going vessel enters a port to load or unload its cargo, barnacles and other organisms can settle on any part of the ship that is below the surface of the water. A luxuriant growth of barnacles on a one-square-foot area of a ship may weigh as much as six pounds. On a large ship, the barnacles and other fouling organisms can add as much as three hundred tons to a ship's weight.

Barnacle fouling of a ship causes a reduction in its speed, increases the amount of fuel it uses, and leads to losses in time and money because of the need to rid the ship of these fouling organisms. The reduced speed of a ship arises when a lot of barnacles settle on the hull and make the ship heavier and its surface rough or irregular. As a result, the ship's engines have to work much harder and a heavily fouled ship may need as much as 50 per cent more fuel to move the same distance it would move with a clean hull. Fouling not only causes fuel to be wasted, but the reduced range of the ship requires more frequent refueling.

Aside from the expense of increased fuel consumption, and wear and tear on the engines, it is costly to dry-dock a ship for

A navigational buoy, heavily fouled by seaweeds, stalked barnacles, and other organisms, requires cleaning before it can be returned to service.

cleaning. Modern day ships are now built of special alloys of steel, nickel, and copper which retard much of the fouling. However, to retard fouling on older ships, we still employ many of the same techniques today that were used centuries ago. When a ship is painted, chemicals such as arsenic, mercury, copper, and zinc are blended in with the paint. The chemicals gradually dissolve in the surrounding waters, and are either toxic or otherwise harmful to the barnacles. However, painting is only a temporary measure because the paint either peels or is worn off, or in time it loses its potency to repel these persistent pests. When we consider that there are probably some 57,000 merchant ships in operation today, then it is not difficult to realize that ship owners spend about one billion dollars every year just to scrape the barnacles off and repaint the hulls with protective paints.

Not only do barnacles slow down a ship, they also reduce the efficiency of the echo and depth sounding equipment, in some cases by as much as 50 per cent. This can lead to hazardous conditions when navigating in shallow waters. Barnacles also clog the waterlines that are used for cooling ships' engines, and they can foul the lines that lead to the hoses used in the fire fighting equipment. This can cause serious problems if the barnacles are not removed and the waterlines checked periodically.

The fouling of buoys by barnacles and other marine organisms requires that these important aids to navigation be removed from the water and cleaned on a regular schedule. Barnacles are indeed persistent and costly pests of the mariner.

Many years ago, A. P. Herbert referred to the infamous reputation of barnacles when he wrote:

> Thousands of barnacles, small and great
> Stick to the jolly old ship of State
> So we mustn't be cross if she seems to crawl—
> It's rather a marvel she goes at all.

8

Barnacle Collecting

Barnacles are fun to collect and study. But we should be conservation minded, and take care not to over collect and especially not to disturb unnecessarily the collecting areas. When turning over rocks in search of barnacles, the collector should try to put the rocks back in the same way that they were found. This is done so other animals, either on the rocks or hiding under them, may resume their normal way of life. We must strive to preserve the natural habitats of the seashore, so others in years to come may enjoy visiting them too. There are other ways to practice conservation. Always put back live animals in the same place if you are not going to save them for your collection. Also, select only a few choice specimens, never take all you find.

When you collect on a private beach or from a private pier, be sure you have permission from the owner before taking any specimens. If you visit a National Park or State Beach, you can collect only dead specimens. You can observe the living animals, but they must not be disturbed in any way. In some states, fishing licenses or permits are required to collect living specimens of invertebrates.

Where to Look for Barnacles

Wherever the sea washes the shore you are likely to find barnacles. They occur on the exposed surfaces of rocks, on jetties and wharf pilings, on lobster pots, and on other man-made objects. If you know where there is a boatyard, you might ask the owner when he plans to clean the hulls and if it would be possible for you to collect some of the barnacles that are scraped off the boats.

Special attention should be given to other living invertebrate animals, because these are a good source for barnacles that settle on their shells and on their spines. Some animals to search for that might be encrusted by barnacles are oysters and other mollusks, crabs, lobsters, sea urchins, sea fans, and corals. We mentioned earlier that turtles, whales, and other sea creatures carry around hitchhiking barnacles. Should you find one of these animals washed up on the shore, you are almost certain to find barnacles attached to them. Beached whales are not commonly encountered. The shells of sea turtles, on the other hand, are frequently found on the beaches of tropical shores, where they are discarded by fishermen. You should also examine the gills of crabs and lobsters where you may find tiny stalked barnacles. The floats from fishing nets, boat buoys, and floating logs washed ashore during storms commonly have stalk and acorn barnacles attached to them.

How to Collect Barnacles

You don't need a lot of equipment to collect barnacles. The well-prepared collector will need the tools and other items mentioned here. The first of these is a small hammer, preferably one that does not weigh too much. Steel chisels, one with a cutting edge of ¼ inch and one with a ½ or ¾ inch cutting edge, are required. The hammer and chisels are used for dislodging acorn barnacles from rocks or other objects. A small awl or pick can also be used for prying specimens from rocks or wooden pilings. When these tools are not in use, be sure to keep the tips covered with a cork or piece of soft wood. A pocket knife can also be

Cecelia Ross searching for barnacles in a tidepool on the Pacific coast of Mexico.

used to remove barnacles from wooden pilings. Work gloves are recommended to protect your hands in case you slip and fall or rub against a colony of barnacles, and they will keep your hands warm in cold weather. You will also need a supply of plastic vials, which you can get from your local drugstore. Small or very fragile dried specimens can be placed in boxes or vials and protected with a piece of cotton or tissue paper. A small notebook and a pencil are essential so that you can record precisely where you find the specimens and the day that you find them. It is a good habit to record the information about the barnacle as soon as you find it. A magnifying glass is also a handy tool to carry with you and, last but not least, a small first aid kit and some food, such as crackers or candy bars. All of your tools and sup-

Giant acorn barnacles, such as Balanus nubilus, *shown here with some of the books biologists use to study barnacles, make good paperweights or pencil holders.*

plies can be carried in a knapsack or any strong and lightweight bag, preferably one that is waterproof and has a shoulder strap.

Before planning a trip to the seashore be sure to check the tide tables in your local newspaper, which are generally in the same section as the weather report. Ideally, you want to time your arrival at the beach when the tide is going out and will soon be at the lowest point. This will give you plenty of time to investigate the exposed shore before the tide starts to rise again. It is not a good idea to collect when the tide is in and at its highest point, because you will not be able to see what you are doing in the water and the surf can be dangerous.

The body of a barnacle will decay and become smelly if not stored in some type of preservative, preferably alcohol, which you can purchase at a drugstore. You can place the barnacle in a vial and add alcohol; 70 per cent rubbing alcohol is fine for this purpose. You can store your specimens in an ice chest or cooler, if one is available, until you get home. Specimens that do not have a body when collected can be stored as they are found, after rinsing them several times in fresh water and letting them dry in the air. Remember that it is especially important to place a label with your collecting information in every vial or box. As your collection grows larger and as time passes, it becomes more difficult to remember the exact date and location of where you collected each and every specimen, and this is why a label should be placed in each vial. When time permits, you can put an identifying number on each specimen or on a little slip of paper that can be placed in the vial or box and on the label too. This is done just in case the specimen becomes separated from its label. The records that you keep in your notebook should also indicate the identifying number of the specimens, because if you lose the label you can always go back to your notebook for the missing information that related to the specimen.

How to Identify Your Specimens

There are many ways you can identify the specimens in your collection, especially if they are not pictured in this book. The easiest way is to get illustrated books describing the marine invertebrates living in the areas where you collected your specimens. These and other helpful books may include some of your barnacles, and may be found in your public library or a nearby college or university library. Many museums also maintain libraries, and these are generally open to the public. A few titles are listed in the bibliography at the back of this book.

There are other means of identifying your specimens and learning their names. One way to identify barnacles is to take them to a museum and compare them with the barnacles that are on display. Another way is to ask the curator in charge of invertebrates at your local museum to help you. If you are near a university, you might ask one of the professors of zoology for help in identifying and naming your specimens.

When searching through books for the names of your specimens, it is important to remember that most books will give you only the scientific names of the barnacles, which are derived from Greek or Latin words. The Latinized scientific name given to each kind of barnacle is a form of universal language used by zoologists. Because each kind of barnacle has only one scientific name generally consisting of two parts, there is little chance of confusing the technical names applied to barnacles. The queen's crown, *Coronula regina*, is the name of a whale barnacle, whereas *Coronula diadema*, another whale barnacle, is commonly called the diadem crown. Few barnacles have been given common names. But if you want, you can consult a Latin or Greek dictionary and try to convert the scientific names into common English names.

How to Arrange Your Collection

There are several ways to arrange your collection, but regardless of the way that you choose, there are certain basic things that you must do to insure that the collection remains of value to you and, if at some time you wish to donate your collection to a museum, is of value to the museum too. First, the specimens must always have an identifying label with such information as the name of the barnacle, where it was collected, when it was collected, and who collected it. At some time you might trade specimens with a friend and the collector's name should appear on his label with the specimens. The more information you can supply with each specimen, the more valuable your collection is to science. Specimens without reliable collecting information generally are of little or no scientific value, unless they are exceedingly rare or unusual specimens.

Scientists around the world today, especially those working in museums with large collections, arrange the barnacles by grouping them systematically. This is done, for example, by putting all of the acorn barnacles in one section of the collection, and all of the stalked barnacles in another section. Of the acorn barnacles, the whale barnacles would go in one place and the turtle barnacles in another, but close by. This is because the turtle and whale barnacles are closely related. The same is true for the sponge barnacles and the boat-shaped acorn barnacles that grow on the stalks and branches of sea fans. In the large group of stalked barnacles, the five-plated forms of *Lepas* would go in one section and the many valved *Pollicipes* in another.

You can store all of the acorn barnacles in one shoe box or cigar box and the stalked barnacles in another. As your collection increases in size, you can arrange your boxes on a bookshelf, or in a filing cabinet or dresser drawers, reserving one drawer for each of the major groups represented in your collection. There are fewer problems concerned with storing dried collections than with specimens preserved in alcohol. The alcohol collections should be stored away from direct sunlight and heat, which cause the alcohol to evaporate more quickly. Alcohol collections should also be stored in a well-ventilated area, and, if at all possible, in a metal container such as a file cabinet.

We believe you will find observing and collecting barnacles to be a most rewarding hobby, one that will enrich your understanding of the complex and exciting world of nature. The next time you are scrambling over the slippery, surf-swept boulders at the seashore, look under your feet; you may well be stepping on the shells of those remarkable animals—the barnacles.

Glossary

acrothoracicans—naked, boring barnacles comprising the order Acrothoracica, meaning "extreme or high chest."

adaptation—the condition of showing fitness for a particular environment; the process by which such fitness is acquired.

annelids—segmented worms, including earthworms, sandworms, leeches, etc.

aperture—an opening, such as that of the shell of a barnacle or mollusk.

appendage—a movable projecting part of the body, such as a leg or cirrus.

arthropods—animals belonging to the phylum Arthropoda, including the crustaceans, insects, spiders, etc.

ascothoracicans—shell-less, parasitic barnacles of the order Ascothoracica, meaning "baglike chest."

bivalve—a clam or other representative of the molluscan class Bivalvia; a shell or covering composed of two valves.

brittle star—an echinoderm related to the starfish.

calcareous—composed of, or containing calcium carbonate; limy.

carapace—a protective shield or body covering of a crab, barnacle, turtle, etc.

chitin—the principle substance forming the outer covering of arthropods and some other animals.

Cirripedia—a class of crustaceans in which the barnacles are placed; the cirripeds.

cirrus—a small, slender, and commonly flexible appendage; plural, cirri.

class—a category in the classification of plants and animals, ranking between a phylum and an order; *e.g.*, class Cirripedia.

commensalism—the association of two or more individuals of different species in which one kind or more is benefited and the others are not harmed.

copepods—members of the crustacean class Copepoda, meaning "oar-footed."
cross-fertilization—the union of an egg cell from one individual with a sperm cell from another; opposite of self-fertilization.
crustaceans—members of the arthropod superclass Crustacea, including barnacles, lobsters, crabs, shrimp, isopods, copepods, etc.
cuticle—a thin external covering on an organism.
cypris—the seventh and last larval stage of barnacles.
diatoms—microscopic algae (marine plants), which secrete shell-like bodies.
dioecious—having the male and female organs in separate individuals; separate sexes.
dugong—a large, plant-eating aquatic mammal related to the manatee; like the manatee, commonly called sea cow.
echinoderms—representatives of the phylum Echinodermata ("spiny-skinned"), including starfish, sea urchins, sea biscuits, sand dollars, sea lilies, etc.
ecology—the relationship of an organism to its environment.
ectoparasite—a parasite that lives on the exterior of its host.
embryo—an animal in the early stages of development before hatching or birth.
endoparasite—a parasite that lives within its host.
exoskeleton—an external supporting structure or covering of arthropods, mollusks, and some other invertebrates.
fertilization—the union of an egg cell and a sperm cell that initiates the process resulting in the development of an embryo.
free-living—not attached or parasitic; capable of independent movement and existence.
gastropod—a snail, nudibranch, slug, or other representative of the molluscan class Gastropoda, meaning "stomach-footed."
gonad—a reproductive organ in which ova (in female) or sperm (in male) are produced.
habitat—the natural or usual dwelling place of an individual or group of organisms.
herbal—a book about plants, expecially medically important herbs.
hermaphrodite—an animal with both male and female reproductive organs.
host—an organism that harbors another.
invertebrate—any animal without a dorsal column of vertebrae.
isopods—members of the crustacean order Isopoda, including whale lice, pill-bugs, etc.
kitchen middens—garbage heaps found at ancient sites occupied by man.

larva—the early stage of an animal, after the embryo, and unlike the adult; plural, larvae.

limy—consisting of, or containing lime or limestone; see calcareous.

manatee—a large, plant-eating aquatic mammal related to the dugong; like the dugong, commonly called sea cow.

mantle—the membrane that encloses the body and secretes the shell of barnacles and mollusks.

marine—pertaining to or living in the sea.

mollusks—animals belonging to the phylum Mollusca (meaning "soft-bodied"), including snails, clams, squids, octopi, etc.

molt—to cast off an outer covering such as the carapace.

monoecious—having both male and female gonads in the same individual; hermaphroditic.

natural selection—the process by which environmental factors eliminate the less well-adapted members of a population, or which causes succeeding generations to adapt to environmental changes; "the survival of the fittest."

nauplius—the first to sixth larval stages of barnacles.

nudibranchs—the sea slugs; shell-less marine gastropods.

operculum—a shelly or horny covering that closes the aperture of barnacles, mollusks, etc.

oral—pertaining to, or being near the mouth.

order—a category in the classification of plants and animals, ranking between a family and a class; *e.g.*, order Thoracica.

organism—a single plant or animal.

ostracods—tiny crustaceans belonging to the class Ostracoda; commonly known as mussel or seed shrimps.

ovum—an egg; the sex cell of a female; plural ova.

parasite—an organism that lives on, or in another at the expense of the host.

phylum—the chief division in the classification of the Animal and Plant kingdoms, *e.g.*, the phylum Arthropoda.

population—a group of individuals belonging to a single species living in a given location at a given time.

predator—an animal that captures or preys upon other animals for its food.

radula—the tooth ribbon of most mollusks.

rhizocephalans—shell-less, parasitic barnacles comprising the order Rhizocephala; commonly known as root-headed barnacles.

sand dollar—an echinoderm related to the sea biscuit and sea urchin.

sea anemone—flowerlike sea animal, related to corals.

sea biscuit—an echinoderm related to the sand dollar, sea urchin, etc.

sea fan—a soft coral; related to the stony corals.
sea urchin—an echinoderm related to the sand dollar and sea biscuit.
self-fertilization—the union of an egg cell and a sperm cell produced by the same individual; see hermaphrodite.
sessile—permanently fixed, sedentary, not free-moving.
sperm—the matured and functional sex cell of a male.
substratum—an ecological term denoting where an organism lives, *e.g.*, rocks, sand, mud, etc.
superclass—a category in the classification of plants and animals ranking between a phylum and a class; *e.g.*, superclass Crustacea.
symmetrical—divisible into equal and opposite parts.
systematics—the science of classification of organisms.
thoracicans—the free-living and commensal barnacles having wholly or partly calcareous shells and comprising the order Thoracica, including non-stalked (*e.g.*, *Balanus*) and stalked (*e.g.*, *Lepas*) forms.
thorax—the major division of an animal next behind the head; the chest region.
vestige—a trace or indication of some organ or part which once existed, but has been lost, especially in the adult stage of an organism.

Selected Bibliography

Barrett, J. H., and C. M. Yonge, 1958, *Collins Pocket Guide to the sea shore.* Collins, London, 272 pp., 40 pls., text figs.

Common marine plants and animals of northern Europe and Atlantic Canada, including 12 species of barnacles.

Bassindale, R., 1964, *British barnacles, with keys and notes for the identification of the species*, Synopses of the British fauna, no. 14, The Linnean Society, London, 68 pp.

Bousfield, E. L., 1960, *Canadian Atlantic shells*, pp. i–v + 1–72, 13 pls., National Museum of Canada, Ottawa.

An illustrated handbook of mollusks of northeastern North America, which also includes 6 species of barnacles, pp. 39, 40, pl. 11.

Cornwall, I. E., 1955, *The barnacles of British Columbia.* British Columbia Provincial Museum, Department of Education, Handbook No. 7, 69 pp.

An illustrated guide to the barnacles found on the Pacific coasts of Canada and the northern part of the United States.

Darwin, C., 1851, *A monograph of the sub-class Cirripedia. The Lepadidae: or, pedunculated cirripedes.* Ray Society, London, pp. i–xi, 1–400 pls. 1–10.

Darwin, C., 1854, *A monograph of the sub-class Cirripedia. The Balanidae, the Verrucidae, etc.* Ray Society, London, pp. i–viii, 1–684 pp, pls. 1–30.

Gosner, K. L., 1971, *Guide to identification of marine and estuarine invertebrates, Cape Hatteras to the Bay of Fundy.* Wiley-Interscience, New York, 693 pp., illus. (Barnacles, pp. 456–462, 7 species figured).

Light, S. F., 1954, *Intertidal invertebrates of the central California coast*, 2nd ed., revised by R. I. Smith & others, 446 pp., 137 figs., Univ. Calif. Press, Berkeley (Barnacles, pp. 128–133, with a key to species).

Pilsbry, H. A., 1907, *The barnacles (Cirripedia) contained in the collec-*

tions of the U.S. National Museum, U.S. National Museum, Bulletin no. 60, pp. i–x, 1–122, pls. 1–11, 36 figs.
Pilsbry, H. A., 1916, *The sessile barnacles (Cirripedia) contained in the collections of the U.S. National Museum; including a monograph of the American species*, U.S. National Museum, Bulletin no. 93, pp. i–xi, pp. 1–366, pls. 1–76, 99 figs.
Ricketts, E. F., & J. Calvin, 1968, *Between Pacific Tides*, 4th ed., revised by J. W. Hedgpeth, 614 pp., 8 pls. (in color), 302 figs. Many references to barnacles.
Schmitt, W. L., 1965, *Crustaceans*, Univ. Michigan Press, Ann Arbor, 204 pp., 75 figs., paperback (Barnacles, pp. 66–76).
Southward, A. J. and D. J. Crisp, 1963, *Catalogue of main marine fouling organisms*. Organization for Economic Cooperation and Development, vol. 1, Barnacles, 46 pp. Available through McGraw-Hill Book Company, 1221 Ave. of the Americas, New York, N.Y. 10020.
Starbird, E. A. and R. F. Sisson, 1973, Friendless squatters of the sea. *National Geographic magazine*, vol. 144, no. 5, (Nov.), pp. 622–633, 10 photographs in color. A well-illustrated article on barnacles.

Index

Boldface numerals refer to the captions to the illustrations.

The Authors

ARNOLD ROSS is Curator of Invertebrate Fossils and Chairman of the Department of Invertebrate Paleontology at the Natural History Museum in San Diego, California.

A native of New York and a graduate of the University of Florida, he is the author of numerous papers and books in the field of fossil and living invertebrates, and is an internationally recognized authority on barnacles.

WILLIAM K. EMERSON is Curator of Mollusks and Chairman of the Department of Living Invertebrates at The American Museum of Natural History in New York City. He is a native of California and received his Ph.D. from the University of California at Berkeley.

Dr. Emerson has contributed more than 100 articles to scientific journals and encyclopedias and is co-author of *Wonders of the World of Shells* and other popular books on shells.